The Experience!

★

The Experience!

How to Wow Your Customers and Create a Passionate Workplace

LIOR ARUSSY

San Francisco and New York

Published by CMP Books
An imprint of CMP Media LLC
Main office: CMP Books, 600 Harrison St., San Francisco, CA 94107 USA
Phone: 415-947-6615; Fax: 415-947-6015
Editorial office: 12 W. 21st St., New York, NY 10010 USA
www.cmpbooks.com
Email: books@cmp.com

ISBN: 1-57820-306-6

For individual orders, and for information on special discounts for quantity orders, please contact:
CMP Books Distribution Center, 6600 Silacci Way, Gilroy, CA 95020
Tel: 1-800-500-6875 or 408-848-3854; Fax: 408-848-5784
Email: cmp@rushorder.com; Web: www.cmpbooks.com

Distributed to the book trade in the U.S. by:
Publishers Group West, 1700 Fourth Street, Berkeley, California 94710

Distributed in Canada by:
Jaguar Book Group, 100 Armstrong Avenue, Georgetown, Ontario M6K 3E7
Canada

Cover and text design by Brad Greene, Greene Design

Printed in the United States of America

02 03 04 05 06 5 4 3 2 1

Dedicated to

Dalya, Cheli, Liad and Netanel with Love

Contents

★

Foreword

The last several years were the "years of the customer." Organizations and executives learned the importance of customers and of their newly found power in the form of the Web. The Internet provided customers with the power to compare and the power to share. These powers allowed them to avoid being hostage to a certain company or brand; customers were in control.

Recognizing the changing rules of the game, organizations rushed to implement customer-loving technologies, all geared to focus on customers. Spending on customer-relationship technologies doubled in the last few years and became a several-billion-dollars-a-year industry. Other related technologies followed suit. The surprising discovery, however, was that, in spite of companies' investment in technology that customers were supposed to love, customers' complaints increased. It seems that we, as customers, did not feel deeply loved as a result of the heavy technology investment; customers were surprisingly unappreciative of all these efforts.

In reality, all these efforts did not take customers' needs into account as much as they focused on maximizing the potential revenues from customers. Companies rushed to reduce costs and automate services, turning relationship into an efficiency machine. The problem is that relationships are not about efficiency, they are about a human touch. One cannot automate relationships. Relationships and experiences can be created only by humans. In the rush to reduce costs and become more efficient, companies mistook improved technologies as the end goal and forgot the human aspects of their businesses. It is through the desire to deliver superior experiences that companies will earn lasting customer loyalty; companies will learn that technologies are merely a tool to reach the end goal. However, the end goal—creating positive customer experiences, one customer at a time—will only be delivered by people. When employees feel fully involved in a company's mission, new technologies can then help them to deliver excellent experiences. Technologies can improve the experiences, but only people can create and deliver them.

Customers are looking for more than just products or the lowest price. In fact, they don't buy products or services anymore, they buy *experiences*. These experiences represent something greater than just the product itself. Customers demand that companies wrap their products in an experience that encompasses excellent service and a delightful attitude toward them, *every time they interact with the company*. Many studies show that the traditional benchmarks of customer loyalty—product selection and price—no longer reign as the reason to prefer a certain company over its competitors. Customers are citing their experiences as the reason for their choice. Customer experience is destined to become the differentiation of the future.

At the core of a lasting relationship is a series of positive, memorable interactions between the company's representative and the customer—those are the positive experiences that must be the company's end goal. This book is about building sustainable relationships through positive experiences. It is a journey of a manager who discovers that people come first and tools come second. This manager realizes that the key to the future customer experience is well rooted in the way old neighborhood stores used to deliver personal service, one customer at a time, one experience at a time. On his journey, the manager discovers the basic unchangeable truths of dealing with, serving, and delivering experiences to customers and employees. He finds his own unique way to develop the experience, a method that will dazzle his company's customers. In the process, he creates an exciting experience both for his people and for himself. Because at the end of the day, customer experience starts with employees' experience, and employees' experience starts with managers' experience. All those experiences are interwoven and connected in an interdependent circle, feeding and strengthening one another.

The strength of the experiences of both internal and external customers will be the competitive strategy for organizations in the years to come. This is a book for all the people who touch customers directly and indirectly. From the top executive to the last customer service representative, we all create experiences, whether we are aware of that fact or not. So if you already create experiences, you might as well create great ones that will bring out the passion in you and will delight and excite both your customers and yourself. Creating and living these experiences is possible. This book is the first step on the path to having an experience of a lifetime, every day.

Let the experience begin.

Lior Arussy
lior_arussy@hotmail.com

Chapter 1

Joe hung up the phone and noticed that his familiar headache was crawling up his neck and intensifying. It was the tenth call that had escalated to him, this morning alone. The customer demanded to know why he was constantly being put on hold, only to end up talking to clueless representatives with a "we don't care" attitude. Like many before him, this caller had demanded to speak to the manager; the representative was more than delighted to send the call to his boss, thereby relieving himself of yet another annoying customer. For what felt like the thousandth time, Joe Jacobs, the call center manager, had to apologize for the behavior, tell the customer that it was an isolated case involving a brand new representative, and promise that it would never happen again.

He was tired of this game. It had been going on for months. He had long since quit bothering to take the customer's information or check the complaint. In the beginning, he was diligent about following up, hoping to solve the problems and improve the service. By now he knew it was useless. His representatives answered the phone with no interest whatsoever—they simply did not care. They now came to work only to earn their money with minimal effort and go home. Customers constantly complained that basic requests were not being answered, and simple issues were taking several calls to resolve. He saw how eager his representatives were to leave for break. Last week he caught a group of them joking during lunch about how stupid their customers were. They were even sharing war stories about how they convinced customers to call at another time.

Joe was embarrassed to be part of all this. This was not the reason he took the job. It was not the reason he joined the call center. He had an ideal of serving people and making them smile. He used to be able to do it when he was an agent himself, but those were the old days. Things had changed, people had changed. His people were different and they simply did not care.

Joe was reviewing last week's performance reports and the trend was clear: more customers placed on hold, more calls escalated or transferred to someone else. It seemed that fewer than 20 percent of the calls were resolved right away. The rest had to wait far too long. The center's ability to resolve customer issues on the first call was decreasing at a frightening pace. As if this were not enough, another ten representatives resigned last week, creating a void—and an additional load on the remaining agents. Turnover was growing, making his hiring and training costs soar. Even more frightening was the fact that the good agents were usually the ones that left. Those agents who stayed either couldn't get better jobs or simply did not want to work hard. No wonder performance was deteriorating at a fast pace.

Just yesterday, he saw a group of his employees in a downtown café. They seemed to be happy and having fun. None of them even bothered to say hello to him. They noticed him, but preferred to ignore him. *How can they be so happy outside the center and so unhappy inside?* he wondered. *Why do they prefer to ignore him? Was he such a bad manager?*

To take his mind off his problems, he turned to his mail. On top of the pile was a brochure from a technology company. He opened it and took a moment to read it. "If your customers are angrier than ever, it is time to treat them well, with our CustomerUno Software. With our technology, your representatives will serve their customers with a smile and deliver excellent customer satisfaction. With CustomerUno," the text continued, "your customers will truly be Numero Uno in your company—they will be treated like kings! We will increase your customer satisfaction by at least 75 percent, or you'll get your money back." The brochure included a set of testimonials from other companies who had turned to CustomerUno to deliver excellent customer service. It seemed that all these companies were happy with their choice. It sounded too good to be true, but Joe was desperate for just a few happy customers. Even if the promises were exaggerated, he mused, he would gladly settle for a 10 percent improvement. He could definitely use a break.

Joe decided to stop being a victim of his miserable situation, and take action to change things. He immediately called the toll free number on the brochure, requesting that a salesperson contact him for a product demonstration and price quote. To his amazement, the agent on the other end was very pleasant and was happy to assist. After asking a few probing questions about Joe's current problems, she immediately connected him with a local sales rep and scheduled a meeting for the next day. In addition, she suggested he visit the product Web site to pick up tips on how other call center managers were benefiting from CustomerUno software. "It will give you some good ideas and show you that you are not alone in these challenges," she said encouragingly.

Joe was impressed. If this call was any indication, he was on the right path. CustomerUno's people seemed very nice, helpful, and caring. He deeply wished his people would be the same. He spent the rest of his afternoon reading customers' stories on CustomerUno's site, happy to see that his challenges were not new or unique. More importantly, he was excited to see that there was a solution out there—a cure for his problems—called CustomerUno.

Chapter 2

Tuesday morning, Joe cleared his schedule and was eagerly looking forward to his meeting with the CustomerUno salesperson. Sleek and professional, Amy walked through Joe's call center problems one by one: high turnover, increasingly dissatisfied customers, low morale, and many others that Joe had not even considered. She seemed to have seen it all before. She spoke about his problems with confidence and knowledge. She explained to Joe how CustomerUno helped a call center turn these problems around and increase customer satisfaction. It seemed like a no-brainer to Joe, so he decided to order the CustomerUno software immediately. Amy offered to expedite installation at no additional cost, to help Joe reap the benefits immediately. Joe thanked her for this gesture.

Joe gathered all his supervisors for an urgent meeting to share the great news.

"Guys, I have some wonderful news for all of you," he declared enthusiastically. "I have just ordered CustomerUno, the number one software tool on the market for customer satisfaction. All the big companies are using it—and it's time we did, too. The company will be investing a large sum of money to acquire and install this technology. I am confident that you will appreciate the importance of this investment." Joe continued, "Starting Thursday, all of us will begin using this software, which has an impressive track record in helping call centers just like ours increase morale and deliver better service to customers."

"You've got to be kidding," was Joan's response. Joan was a good supervisor and her team was one of the better performing teams in the center. She usually had a positive attitude and an open mind toward changes and improvements. She often came up with suggestions of her own about how to make things better. Her response caught him by surprise. "The last thing we need right now is to learn another product," she continued. "The last software we installed was a complete disaster and delivered nothing but cynicism and complaints, not to mention wasting our time. None of us needs another one of those technologies. We've already got plenty of problems as it is."

He noticed many nods in the audience. It seemed her views were shared by others, but Joe was determined not to let this resistance stop him from leading his team on the path to success. Joe decided that they were simply too preoccupied with the old way of doing business. They simply did not know what was good for them. He decided he would lead them toward the solution and prove them wrong.

"Well, Joan, it is all in the attitude we share with our employees," Joe responded. "If you're excited, so are they," he said, "and I expect you to be excited about CustomerUno and to share that excitement with your staff." He was proud of himself. His voice commanded leadership and determination, just what was needed in this situation.

"Well, why didn't you share it with us before deciding for us?" The question came from Kevin, the ultimate complainer. Kevin was the kind of person who found fault with everything. Kevin had worked at the center for years and was a fixture. He always saw the negative in any situation and was always looking for a reason to avoid change. Joe did not expect a different approach from him now.

"First, it never occurred to me that you would oppose a solution that can help us increase morale. We all stand to benefit from it. Second, there are moments when a leader needs to make decisions alone—and this is one of them," Joe concluded with conviction.

He could see in their faces that his response only increased their reluctance. They definitely did not share his enthusiasm, but he was determined not to let this initial response tamper with his solution. Joe was getting more confident about his choice and was determined not to let the criticism drag him down.

While they were leaving the room, he caught Mark saying to Joan, "Look at the bright side, it could have been worse: he could have told us that we actually need to be nice to customers. This way it's the job of that new software, not ours." It was one of Mark's endless chain of cynical jokes. Everything was potentially a source of sarcastic remarks for Mark. Joe decided to ignore him as well.

Joe returned to his office and composed a memo:

Dear Staff,

This coming Thursday marks great progress for our call center. On Thursday we will be implementing CustomerUno software. This software is responsible for the success of many leading companies. CustomerUno allowed those companies to deliver excellent customer service and increased customer satisfaction by as much as 75 percent. I am confident that you all know the importance of customer satisfaction to our company. Therefore, I am sure you will all appreciate the investment that we are making to improve our customer service through this exciting new technology.

Throughout the day, CustomerUno's experts will install the software in all the computers in our call centers, so that we can benefit from the technology immediately.

I thank you all for your cooperation. I look forward to stellar customer service as a result of this new software.

Let us all make our next customer Numero Uno!

—Joe Jacobs

Joe was very impressed with his e-mail. The tone was positive, just what his people needed. He especially liked the caring touch he added at the end of the message. He was sure that they would get it. He rushed to send it to all his employees, making sure to copy his boss, as well. A little bit of self-promotion never hurt anyone.

CustomerUno's staff worked aggressively, and by the end of the day on Thursday—as promised—the software was installed on all the center's computers.

The implementation was going well, with no interruptions, which Joe took as a good sign. His e-mail must have helped people understand the importance of the new technology. He was looking forward to the results.

Chapter 3

"Hello, to whom am I speaking?" asked the angry voice, on the following Monday morning. It was one of those endless escalated calls that was transferred to Joe every day. What a great way to start the day, he was thinking cynically to himself.

"This is Joe Jacobs, call center manager," Joe replied calmly. He was used to the routine by now.

"Well, this is your CEO and I want to tell you that your service is lousy, the worst I have ever seen. I am ashamed that the people who just answered me are actually responding to our customers. If I were a customer, I would have defected to the competition before you could hang up the phone." Joe was not ready for this.

"But can you please tell me what the problem is?" Joe was terrified. This was not supposed to happen.

When the new software was installed, quality of serv-ice should have improved.

"Let me tell you what the problem is," the CEO continued yelling. "I called to test your center and asked a simple question about our new product. The agent had no idea what product I was talking about, even though I actually spelled the name for him. Then he put me on hold for ten minutes before transferring me to another agent, who told me that the product does not exist, and that I should check my facts again before calling next time."

"This is an isolated case, sir. If you tell me who you spoke to, I can guarantee you that the person will be fired, sir," Joe replied.

"I do not think that this was an isolated case. I would like you to guarantee me that all your agents will be knowledgeable and will have a better attitude next time I call. Joe, you've got about a week to fix this situation," the CEO concluded, and hung up before Joe was able to respond.

This was the last thing he needed right now. He was surprised to hear from the CEO. He hardly knew what the CEO looked like. The call center was placed in a bad part of town, some one hundred miles away from company headquarters. He never heard from the people at HQ unless the budget needed to be cut. No one ever asked him how his center was doing or asked for any reports. Occasionally, an upset customer managed to get to some executive, and then Joe would get the usual call from his boss, asking him to make sure that all customer problems stayed contained in the call center.

"That is your job," his boss would remind him. "Your center is here to make sure that no one in top management—nor anyone else in the company, for that matter—will be bothered by customers." His people were playing a very important role in the company. They were the gatekeepers who enabled management to peacefully ignore their customers.

So why did the CEO call the center? It was strange.

Joe reviewed the results of last week's performance. It looked the same as always. More agents left the center last week, more calls were transferred from one agent to another, and customer satisfaction was showing the same declining trend. The summaries and graphs behaved as if CustomerUno had never been installed. He knew he had to find the heart of the problem.

Frustrated, he decided to walk around his center. Everyone seemed to be busy talking on the phone with customers. The electronic tracking board showed that over one thousand calls had been handled so far today, with twenty-three calls waiting to be answered. The average wait time was seven minutes. "I would never allow anyone to put me on hold for seven minutes," Joe thought to himself. "That is just way too long."

"I sincerely understand what you are saying, sir. Let me check to see what I can do to help you. May I put you on hold for a minute?" It was a new voice, probably one of the most recent recruits. They come and go so quickly that Joe had stopped keeping track. The nameplate on her cubicle said that her name was Dawn. Her cubicle was neatly organized, and contained many personal pictures and keepsakes.

"Well, sir, I have some good news for you. I have just arranged for a pickup of the defective product from your house, at no charge to you. To avoid any inconvenience, I also arranged for an immediate replacement to be shipped to you. Would that be to your satisfaction?" Dawn paused for his answer. Her smile confirmed that the customer was satisfied. "Good, sir, I'm glad to have helped. I would like to again apologize on behalf of the company for the trouble this problem caused you. Have a wonderful day." Dawn ended the call with a big smile on her face.

Joe was shocked. He couldn't recall such a good

attitude in his center for quite some time.

He introduced himself. “Hi, my name is Joe Jacobs, and I’m the center’s manager. Do you mind if I spend some time listening to some of the calls you receive?”

“Hi there, I’m Dawn,” she replied. “Feel free to listen.”

Joe spent the afternoon listening to the Dawn’s calls. They were all pretty typical: upset customers with similar complaints and frustrations. Yet she handled each of them in the same manner as she’d handled the first call he’d heard. She was happy to help. She acted as if it were a special privilege for her to be of service. It seemed as if she really enjoyed her work. If it was needed, she would navigate the company’s Web site for answers, or call others and ask them for help, but she never sent the customer to someone else. Dawn made sure that customers were actually satisfied with her proposed solutions, and always looked for something extra to offer them to make sure they were truly delighted. All the customers she handled were very happy with her approach and her solutions, and they complimented her accordingly.

It was like a spot of light in a sea of darkness. At the end of her shift, he invited Dawn for a cup of coffee at the local café.

"How do you do that?" he asked. "I just checked your records, and it seems as if this day was no exception. Your customers are always satisfied and you hardly ever transfer calls to other agents."

"Well, I just enjoy helping people," Dawn replied. "I like being able to solve their problems and make them happy. That's what makes me happy," she added.

"You must have used the new CustomerUno software, right?" Joe asked.

"What software are you talking about?" Dawn asked.

"The new software we installed last week to increase customer satisfaction."

"Oh yes, I remember now—the e-mail you sent, in which you told us that the technology will serve customers better. No, actually, I'm not using it at all. I don't even know how to use it. I don't believe in technology anyway. I believe in people. I've got to go pick up my daughter from the babysitter. Nice chatting with you."

As she was walking away, he realized that he had made a huge mistake. He took his notebook and wrote, in big letters, the first principle:

Technology does not care for customers, people do.

Technology
does not care
for customers,
people do.

It dawned on him how mistaken he'd been. The software was not his real problem. His people didn't care—that was the heart of the problem.

Slapping the technology on them did not make things better. In fact, it actually made things worse. The idea that technology would serve people better was an illusion. He went for the quick fix, losing sight of the real problem: his people.

While driving home, Joe was on the phone with his credit card company; he was trying to sort out a charge that was made to his account in Lima, Peru. He had never been in Lima, and he disputed the charge. Joe had already called three times regarding the same problem and had submitted three different forms. The process had taken over a month so far, and no resolution was on the horizon. He was really getting angry by now. As he placed his call, he was put through the same annoying automatic system that gave him endless options to obtain information he was not interested in. Frustrated, he had no alternative but to follow the menu until he reached a live agent. The first agent he spoke to told him that his paperwork had never reached the company, and it was only after ten additional minutes of holding that someone actually found it and promised to call him back. This was the third time the company had promised him the same thing, yet no one ever bothered to call him back. He did not know what to do anymore. It seemed as if no one cared at his credit card company. The agents were fielding calls as if they were in a factory, and each customer was a nameless product that needed to be moved through the process quickly.

Joe was sure that his credit card company had plenty of technology installed in its call center, yet the service was no better than it was at his center. It actually sounded much worse. He reflected on his realization. The answer to his problem lay in the people, he thought, not in the processes or technologies. He decided that if the answer to the problem was in the people, then they needed to be part of the process, and he shouldn't make decisions for them.

His personal experience with the credit card company helped him recognize the problems he had at the center. Now he had some great ideas for his discussion with his team tomorrow.

Chapter 4

All the way to the call center, Joe was rehearsing his opening statement for the meeting with his supervisors. He was not sure of the ultimate outcome of the meeting, only how he was going to open. This meeting should be different than previous meetings, he thought.

While his supervisors were gathering in the room for the regular morning meeting, he reviewed, one last time, his opening statement.

"The other day, I got a phone call from our CEO, and he was upset. He pretty much notified me that our low performance level and our poor attitude toward customers would no longer be tolerated. He gave us a week to fix things around here," Joe opened.

The silence was complete. No one uttered a word as the weight of his message settled in. They had never met the CEO and never thought he knew who they were or what they did. This message from him was completely unexpected.

"After thinking things through, I also realized that I made a mistake, and that you were right. Technology is not the solution. Technology is merely a tool. If the users will not take advantage of it, it is useless. I was wrong to force it on you. I truly hoped that it would cure our problems." As he was speaking, he noticed different expressions on the faces of his people. Cynicism was replaced by a mixture of surprise, puzzlement, and sympathy. Yes, for the first time he saw sympathy on their faces. It's a good sign, he said to himself, and continued.

"I must tell you, that it was actually Mark who taught me the biggest lesson." Joe noticed that Mark was shocked to hear his name associated with the solution. The rest of the group were carefully listening now to see what he would say. None of them could imagine that Mark was the key to the solution. He was, after all, just a harmless jester.

"The other day, when I announced the new software installation, as you left the room, Mark commented to Joan that the meeting could have been worse. He said that the worst was asking people to actually deliver better service. The truth is, that's exactly what I should have been asking for. There is no other way.

"The biggest realization I had, though, is that the people, and nothing else, are the key to our solution. If people want to, we can change things pretty quickly, but if they don't, then we might as well give up and go home. No new technology, no enhanced procedures, can make our people smile. No imposed-from-above process will make them be nice and deliver a pleasant experience. To achieve that we need to go further than our regular management tools. What we need is beyond the official rules, performance evaluations, or even monetary bonuses. We must reach deeper and build an environment that will foster smiles and a nice attitude toward our customers. We must create an atmosphere of caring. I would like your help in building and implementing the plan to change this place and deliver top-notch service." He noticed a few nods of agreement in the group and he felt good about most of his listeners' responses so far. However, some were still hesitating. "This time I am not going to do it alone," Joe concluded.

"To start the process, let's do a little exercise. I'd like you all to write down the best experience you've ever had as a customer. Please take few moments to think about it." The room became silent as they each reflected on their best customer experience.

"Let's share our experiences. Joan, why won't you go first," Joe suggested. He was hopeful, but a little nervous. After her comments during the last meeting, he was no longer sure if he could count on Joan's support.

"My best experience as a customer was during my last vacation," Joan offered. "We drove down to Orlando and when we showed up at the hotel, we were told that they were completely sold out and had no rooms for us. They did not have any record of our reservation, even though I secured it with a credit card. Our complaints didn't help, since the hotel simply did not have an extra room to spare. Angrily, I called the hotel chain's toll free number and started to yell at the first agent who answered me. Instead of being upset, the agent on the other end of the line was calm and asked me to explain the situation to her. In a matter of minutes she located another hotel two miles away and arranged a new reservation for us. As a form of compensation she got us upgraded to a suite and had a nice basket of fruit and a bottle of wine waiting for us."

"What made this experience an example of good customer service?" Joe asked.

"The agent did not have to go the extra mile for us,

and yet she did. She recognized how devastating it can be to start your vacation like that, and she didn't want this bad experience to cast a shadow over the rest of our trip."

"I guess what she did was to simply treat you the way she would have liked to be treated," Mark said—this time without any touch of cynicism.

"Exactly!" Joan agreed.

Joe summarized, "When we wake up in the morning and brush our teeth, we are customers of the toothpaste company. We get dressed and we're customers of the clothing manufacturer. We eat breakfast and we're customers of the cereal and coffee companies. We drive to work and we're customers of the car manufacturer. At work we use phones and computers, so we're customers of the phone and computer companies. As customers, we demand flawless service and excellent quality. We will not tolerate less than that, and we definitely will reject any rude service we get from those companies. It's our money and we demand to get its full worth."

They all nodded in agreement, even though they weren't sure where he was heading. Joe smiled to himself. He realized that his personal experience as a customer was key to his understanding his own customers. We're all customers, after all.

"So how come, when we are talking to our customers, we don't understand their expectations?" Joe asked. "How come we cannot relate to them and deliver to them the same kind of service we would expect as customers? How come we forget what it is to be a customer the moment we find ourselves on the other side of the phone? We are still the same people," Joe concluded.

The silence was complete. They all sat quietly; no one said a word. It sounded something like one of those quarterly "Do a better job ASAP" speeches that they'd heard dozens of times before, except that this one sounded different, somehow. It actually made sense. They knew that Joe had a point.

"That is the cornerstone of the change we need to initiate here together," Joe said. "Treating others the way we want to be treated. We are all customers of other companies, and we know very well how we hate being mistreated. We sure do know bad service when we experience it ourselves. Let's treat our customers the way we want to be treated."

Joe wrote on the board in large letters the next key principle:

Treat the customer the way you want to be treated.

Treat the customer
the way
you want
to be treated.

One after another they all shared their best experiences, stories about someone treating them well because that was how that person would have loved to be treated. It was inspiring to hear about customer service agents at other companies who actually took the time to help people they didn't know. Each of the stories carried a different message and left them with food for thought and with ideas to implement.

"So what are the characteristics of a good experience?" Joe asked. "What do all good experiences have in common?" The supervisors all contributed their ideas and Joe wrote them on the board:

- Passionate people
- Positive attitude
- Listening
- Caring
- Willingness to help
- Sincere desire to understand
- Human interaction
- Respect
- Fun
- Pleasant outcome
- Excitement

"These are all correct and that is what we all must achieve. Let's take this exercise and share it with our teams. I want each of you to take an hour today to run the same questions with your team. Hopefully, we will all come to the same conclusion. Let's meet tomorrow to continue our discussion." They all nodded in agreement, but he thought he could sense more than just agreement. It looked as though they were actually willing to give it a shot. It seemed that they were all tired of the old way of doing things.

"As you're working with your teams, please try to clarify with them the expectations of our customers. I suggest that you list the most common complaints we get and ask your teams how they think the customers want us to respond. Afterward, let's see what we need to do to make it happen," Joe concluded. He handed each of them a piece of paper with the following table:

Customers' complaints	Customer's desired resolution	What we need to achieve it
1.		
2.		
3.		
4.		
5.		

They all took copies of the table on their way out of the meeting room. Joe noticed a different attitude in some of them. They left the room a bit more willing to cooperate and more willing to give it a chance. He was hopeful, but he also knew that success was no longer his personal issue: Without his people he would fail. He must turn them into believers, so they in return would turn their people into believers. This would need to be a joint effort. His people would have to share both the sweat and the glory. It was not just his success on the line, it was theirs as well. They were all in it together. He truly hoped they understood that.

Joe spent the rest of the day walking around and chatting with various agents on the floor. He liked this new routine he was starting. In the past he had always been busy handling his reports, the new recruits, and the escalated calls of upset customers. Time was simply not available for walking around. But today he forced himself to drop everything and just get a feel for what was going on. He was sure that the rest of the problems would wait for tomorrow. Joe noticed many new faces out there. Many of them did not even know him. It was actually a very pleasant experience talking to them. All of a sudden, they were no longer statistics or components of a pie chart, but real human beings.

They had plenty to share with him. They seemed to appreciate the time he took to talk to them.

He stopped by Dawn's cubicle and thanked her for last night's discussion.

"No problem, it was my pleasure," she replied with a big smile. "Anytime."

"There is just one question that's still bothering me. I noticed that you never transferred any call to another agent. Even though you're new here and don't know all our products by heart, you consistently insisted on handling those calls, even though you could easily have transferred them to the experts group. Why didn't you?"

Dawn hesitated for a moment and then replied, "First of all, I hate it when others do that to me. So I try my best not to do it to others. It's simple common sense. But in addition to that, whether you're at work or on your own time, you can't throw your problems on someone else. You have to deal with them yourself. These are my calls and I have to deal with them."

"Thank you for the insight," Joe said as he continued his walk around the center.

As he continued, he noticed that all the supervisors had done what he'd asked them to do in the morning meeting: they were conducting small meetings with their teams. He was worried that they wouldn't be able to deliver the session the same way he had, but he knew that he had to trust them and couldn't do the work for them. The key to success was their full participation.

As he drove home, he thought more about what Dawn had told him that day, and it gave him an idea for the next day's meeting.

Chapter 5

This time as he entered the meeting room, the supervisors were all there waiting for him.

"So how did it go yesterday? What reactions did you get from your teams?" he asked.

"My team reached the same conclusion that we did in our session," Joan said. Others nodded in agreement. They shared some of the notes they took from their sessions and it seemed that they had all reached the same conclusions.

"In that case, if we already agree about what constitutes a good customer experience, let's spend some time talking about your worst customer experiences," he said.

"Well that's an easy one, I've got plenty of those," Mark said. They all smiled. If Mark was volunteering, it was sure going to be entertaining.

"Go ahead, Mark," Joe said.

"Mine just happened five weeks ago. I ordered a gift for my son's birthday through a Web site. The gift didn't show up in time for his birthday. I had to go and buy him another gift at the last minute and spend more money than I had originally planned. When the original gift arrived and I tried to return it, the agent from the Web site refused to accept it, claiming it was the shipping company's problem, not theirs. He advised me to go to the shipping company and ask them for compensation.

"Needless to say, the shipping company refused to deal with it either. I then called the Web site again and asked to return the gift. I was put on hold for an eternity before the representative transferred me to another agent. That agent listened to my story and claimed that it did not make sense at all. He then put me on hold again before transferring me to a third agent, who claimed to be the supervisor. This third person told me that I needed to submit my request in writing, with full details of the incident, so they could consider whether or not to accept the return. It's now five weeks later and I still haven't gotten an answer from them."

"Now that is just plain rude!" Joan exclaimed.

"I can't believe companies like that are still in business," Kevin added.

"Some of our customers would claim that our company is so bad they can't believe we are still in business," Joe told everyone. "What do you think is the main problem with the service that Mark received?"

"No one was willing to take responsibility," Joan said.

"That's right," Joe replied. "We've got a serious problem here of ownership and responsibility. The agents think that since they did not create the problem, they don't need to solve it. They basically spend their time transferring problems to someone else or trying to convince the customer that it's his problem, not theirs."

Joe then went on to talk about his discussion with Dawn, with particular emphasis on her view that in every aspect of our lives—whether we're at work, at home, on the road, or wherever—we need to handle our responsibilities, not hand them off to someone else. The group seemed impressed by the insight. It made complete sense. Joe was proud to have someone like Dawn working in his center.

He wrote on the board the next principle:

Own the problem.

Own the problem.

As the group shared some of their worst experiences as customers, they noticed that all of the examples had something in common: a lack of ownership and responsibility on the part of the agents and the companies they represented. The agents viewed themselves as problem avoiders, not problem-solvers. They always tried to shift the problem to someone else or blame the problem on another person. Sometimes, when all excuses were exhausted, they simply blamed it on the customers. Everyone in the room clearly realized that this was the wrong approach.

"What are the characteristics of a bad experience?" Joe asked.

The group listed them on the board:

- Agents who do not care
- Lack of ownership and responsibility
- Lack of interest in solving the problem
- A focus on the process, not the customer
- Bad attitude

"Let's take this exercise to our people and talk to them about taking ownership and responsibility, even when we did not create the problem," Joe said, concluding the meeting.

As the supervisors went back to their offices, he noticed that they were making progress. The cynicism that used to accompany every meeting had been replaced with relevant follow-up discussion. It looked as if they were truly trying to change.

"I disagree completely with what you said!"

It was Tim, a longtime agent, during Kevin's team meeting later that day.

"I will not take responsibility for other people's mistakes and wrongdoing. All that we do here—all day—is take care of people who are yelling and screaming at us for mistakes made by others in this company. As far as I'm concerned, your suggestion that we should own the problem implies that we created it. It's like we're relieving others in the company of their responsibilities," Tim argued. "I did not decide to manufacture a defective product. Let the CEO who approved the decision bear the consequences. Why is he not here to talk to customers and own HIS problems, problems that are the direct result of HIS mistakes?"

Others nodded in agreement. It seemed to be simple common sense. Kevin was not sure what to say to this argument. Tim's suggestions made sense to him, too. The CEO and the rest of the executives should be staffing the call center and taking a little responsibility for their own mistakes. Now that he thought about it, it wasn't only a matter of bearing responsibility: Everyone in the company needed to listen and create a wonderful experience for customers. It didn't make sense that the whole burden should be on the call center agents. The best customer experience is the cornerstone for the whole company's success. If we can all give the customers a great experience, the company will win; if we can't, nothing else will matter. Even if all of the call center agents changed their attitudes, they are only part of the overall experience. Kevin decided to bring the matter up with Joe and see what he could do about it.

"Tim, you've got a point that I didn't see. It does make sense. Let me talk to Joe and see what I can do about it. What you're suggesting is not a trivial change, but it does make sense," Kevin replied. He noticed that a few of the faces around him registered surprise at his response. They expected him to argue with Tim and ultimately disregard his view as unrealistic or irrelevant. In the past, their opinions and suggestions usually had not gone anywhere. No one ever asked them for their ideas or did anything with the suggestions they provided.

Chapter 6

"Joe, we've got to talk. Do you have a minute?" Kevin asked, while entering Joe's office.

"Yes, come in. What's up?" Joe answered.

"Well, you know the session I did today about customer experiences and ownership? To my surprise, I got some strong resistance from my team, and I must admit that after thinking about it, I think they're right."

"What are you trying to say? That we don't need to be responsible for the problems?" Joe asked, surprised. He liked Kevin and thought he was a reasonable guy. This argument was completely unexpected.

"No, that's not it. We should own the problems coming our way. But we must also involve the people who created those problems. Otherwise, we're owning problems that other people created, and we're relieving them of their responsibility for their actions. If a faulty product upset our customers, then the engineers, product designers, or production people should bear the responsibility. Actually, without the whole company being involved, the experience we create will simply be an isolated incident, and the customer will never be fully happy with our company. We'll deliver one kind of experience, while other parts of the company deliver a different one. The lack of consistency will ultimately confuse customers, not delight them."

"So what do you suggest?" asked Joe.

"I suggest that the CEO and his executive team join us here in the call center, answering calls and listening to the consequences of their wrong decisions and faulty operations. Without doing so they will never fix the problems," Kevin said.

Initially, the idea sounded strange. After all, Joe did not even know the CEO. His only encounter with him had been so negative that he didn't want to speak to him ever again. The executive team resided at headquarters, about one hundred miles away at a prestigious business park. He often felt that the people at HQ distanced themselves deliberately, in order to avoid the call center.

The more Joe thought about it, the more he realized that Kevin was right. It seemed that the real distance between the call center and headquarters was much greater than the actual physical distance between the two buildings. After further discussion with Kevin, he realized that he had to try to involve the entire company. It would be wrong to assume responsibility alone. Joe decided to take action.

Joe placed a call to his boss and asked for permission to approach the CEO. "If you want to get yourself in trouble, go ahead. I'm not getting involved," was the reply he got. Assuming he had nothing to lose, he picked up the phone and called the CEO's office.

"Who may I say is calling?" was the question on the other end of the line.

"Joe Jacobs, the manager of the company's call center," he replied.

"Do you have a scheduled appointment?" he was asked by the same firm, superficially kind voice of a professional gatekeeper.

"No, I do not, but it is an urgent matter," Joe replied.

"What is it regarding?" The voice asked, looking for any excuse to get him off the phone as soon as possible.

"It is about his direct responsibility to our customers and the future of their loyalty to this company," Joe answered, trying very hard not to get annoyed.

"Please hold," the voice replied.

After few minutes—that seemed like eternity—a grouchy voice responded.

"What can I do for you?" the CEO asked.

"Well, sir, last week you called me, accusing the call center of mistreating our customers. Your claims had some merit, so I have investigated the matter and made some changes, but we do need your participation. You see, our agents seem to have difficulty with the fact that corporate makes decisions that produce faulty products, yet the people who make those decisions are not there to take responsibility for the results of their actions."

"So what do you want the agents to do, run the company—or are you suggesting that we should all be fired? Tell your people to do their job and that's it," the CEO said testily.

"No, sir. Actually, what I suggest is that you and your executive team join us occasionally and talk to customers. Doing so will give you a better idea of what's going on in the minds of customers, and it might influence your decisions and actions on their behalf. In addition, our agents will see you personally responding to customers and will better appreciate how important their role is in the company. They are more likely to take their jobs seriously then, and to deliver exactly the results you're asking me to deliver. After all, serving the customers is the job of everyone in the company, not just the task of the agents in the call center. We are all responsible for delivering the best customer experience."

There was silence on the other end of the line. Making this call was a huge risk on Joe's part. The butterflies in his stomach were clearly in action. He could almost hear the CEO's brain processing his last message. Joe knew that this was a moment of truth. It could either be a huge victory or a total disaster. Judging from the way the conversation had gone thus far, the disaster was about to arrive. Still, a voice inside Joe convinced him that no matter what, he was doing the right thing. It simply made sense. The alternative was wrong.

"You've got a point there," the CEO said. Joe breathed a sigh of relief. Joe explained to the CEO the steps they had taken so far, including getting the full participation of everyone in the center in order to design their new guiding principles. He described to the CEO the principle of problem ownership. He recited some of the factors that his staff had pinpointed as vital to creating the ultimate customer experience—an experience that could only be delivered if employees were to serve others as they themselves wanted to be served. The CEO realized that Joe's request was motivated by a genuine desire to improve customer service, and that what he was proposing could bring about a real solution.

"You know what, tomorrow I was planning a golf retreat for our executive team. I think we'll make a slight change in schedule and come to visit you guys."

Joe was excited. This was starting to work out. He called Kevin and informed him of the good news. Kevin suggested that they prepare the center for the special visit.

The next day, when the executive team arrived, a huge sign welcomed them to the call center. The sign read:

Technology does not care for customers, people do.

Treat the customer the way you want to be treated.

Own the problem.

The customer is everyone's job!

The customer
is everyone's job!

The last principle had been written by Joe and his team during the previous night's meeting. They all agreed with Kevin's idea and applauded Joe for taking the risk to present it to the CEO. They all shared the excitement of having the executive team visit the center. They decided not to prepare their people, but to let them mingle and speak to the executive team as openly as possible.

Joe escorted the executive team to the meeting room, where the supervisors were already waiting for them. Joe outlined the procedures of the center to the executive team, then suggested that they spend the day answering calls. The executives were quite surprised at the suggestion. Not only had their golf outing been canceled, now they were expected to do this low-level job. Joe handed each executive a brand new headset engraved with his or her name. "We hope these will be used more than just today to serve our customers," Joe said.

The CEO intervened: "Yesterday Joe called me, and he had a point about something we have all neglected for too long. We are all making decisions every day that affect our customers, yet we never confront the consequences. We always assume that these agents will bear the bad news associated with our bad decisions. Well, it's time to forge a link between the decision-makers and the customers. If we made bad decisions that resulted in bad customer experiences, then we need to face the consequences of those choices. We can no longer let others do it for us."

"So what are you saying—that from now on, it's more important for us to waste our time answering phone calls?" asked the Vice President of Research.

"Let me ask you a question," the CEO responded. "If you designed a faulty product, why do you think that Dawn, our agent here, should be getting an earful of complaints from disgruntled customers, who will yell and scream at her, even though you are responsible for the bad design and will go on with your life as if nothing happened?"

The VP of Research wisely chose to keep silent.

"What I am saying," the CEO continued, "is that we are all responsible for the customer experience. If we don't participate in the process of actually talking to customers—and listening to them—we are doing our job in an irresponsible way. And yes, I think this exercise today should be conducted more often and by everyone in the company. We are all responsible for our customers' experiences."

A few heads nodded in agreement. It looked as if

the executives were getting the point. Joe assigned them each to a supervisor who would mentor them through their shifts.

The excitement was quite high among the agents, who saw corporate executives taking calls. It was the first time they felt as though someone at corporate actually cared. During lunch, Joe invited the executives to mingle with the agents and talk to them. He could see the excitement in the executives' eyes as they shared the details of the calls they took during the morning. The agents offered ideas from customer calls they had taken over the last few days. It was a productive lunch, and everyone seemed excited.

Following the afternoon shift, the executives gathered again in the meeting room with Joe and his team of supervisors.

"Wow! That was a grueling experience talking to upset customers all day," said the Vice President of Operations.

"It's our job," said Joe, "and we're proud of it. But you can definitely make it easier, if you'll take what you learned here and improve the products, so that we'll have fewer upset customers."

"I actually picked up a couple of ideas from customers today," said the VP of Research. "While talking to them about our products, I asked some questions about ideas we're toying with in the lab. They were very happy to share their opinions and suggestions. It was a great experience for me. I'm definitely going to come back again soon," he added.

"I just learned about some of our competitors' promotions today, and I'll be able to respond quickly to improve our competitive position," the CEO said. "As far as I'm concerned, today was not a onetime event, but a beginning of a new way of dealing with our customers. From now on, we are all owners of the customer experience. The customers create our brand, and their loyalty is the strength of this company. We are all responsible for it. Anyone in a decision-making position in the company must staff the center regularly, so they can be connected with the customers, gain immediate knowledge of their needs, and share responsibility for delivering the best service possible. To make things easier, I have decided to open a satellite call center in our headquarters, so you can all visit it regularly, check with the agents, and find out what's new with our customers," the CEO concluded.

"Joe, I want to thank you for this excellent idea. We have gained valuable information today, but more importantly we learned about our role and our responsibility in the relationship with our customers. We can no longer delegate the responsibility of taking care of our customers to one group of people in the organization. The customers are everyone's business. If anyone in this company is doing something other than caring for the customers, they're doing the wrong thing. We'll be back here more often," the CEO promised.

"Well, the thanks for this idea doesn't belong to me," Joe replied. "It was actually an idea generated during our sessions with our agents. The person who came up with it is Tim, one of our best agents," Joe concluded. He introduced Tim to the executive committee, who all thanked him for his willingness to share his ideas and help the company improve.

Later that afternoon, Joe was reviewing the reports of the center's performance. It was the first time he noticed a positive change. Customer surveys showed a growth in satisfaction compared to last week's results. The agents seemed able to answer questions more quickly and to resolve issues faster and more efficiently. He realized, though, that he didn't feel he needed the reports to tell him that improvement was coming. As he walked around the center and chatted with the agents, he noticed a positive difference.

Some of them actually smiled and showed some energy. He decided to hit the floor again. It seemed he was doing a better job there than in his office.

Chapter 7

"We are in the business of unpredictability," Joe started. This was one of his recent initiatives—he was having a town hall meeting with his agents. He decided that they needed something in addition to the messages they got from their supervisors; they needed him to communicate with them and to listen to them firsthand. "We never know what the next call will bring our way. A happy customer or an upset one. An easy-to-solve problem or an impossible problem. And within this environment we're working our hardest to deliver delightful experiences to our customers." Joe continued, "How many of you received a call in the last week from a customer who was upset but really didn't know what he wanted, and half of what he said didn't make sense?" The majority of the hands were raised, and agents were smiling while recalling those calls.

"It happens all the time. Why do you think it happens?" Joe asked.

"Because they got moods," replied one of the new agents, while everyone chuckled in recognition.

"That's right," Joe said. "They're not after us or trying to hurt us personally. They're simply human. As human beings they have moods and sometimes when things are really tough they can't take it any longer and they take it out on the first person they come in contact with. Often that person is one of us."

They all nodded in agreement. He felt that a connection was forming with his agents. And not just because of this open discussion. The connection had been strengthening over the last few weeks.

"I'm sure that in your personal lives you often encounter similar behavior; a spouse or close friend may lose their temper, blame us for something we didn't do, or simply make us feel bad, for no apparent reason. How do you respond to such incidents in your personal lives?" Joe asked them.

"I just ignore it," Tim said.

"I try to make them feel good by doing something they like. I hope that the good I'm doing will somewhat offset the bad feelings they have," Joan said.

"I always remind myself that it's not personal, even

though it sure sounds like it. I remind myself that either their stress or their anger about other things leads them to this behavior," said Dawn.

"These are all great ideas. What sound does a smile make?" Joe asked, causing them all to wonder where he was going next. "Even though a smile does not make a sound, you can recognize a smiling person by the lilt of his or her voice. The best way to handle difficult situations is by first remembering that customers are human beings. Their emotions can make them less reasonable than they normally would be, and they might blame us for things we don't have control over. Relating to our customers as fellow human beings—in a way that takes us beyond just their names in the database—can help us help them. The next step is to use that same connection to help them see the problem in perspective and start working on an agreed-upon solution. Smile at them. Make a human gesture that brightens their day in some small way. Show them someone out there cares about them. They might not be able to see you smile, but they sure will recognize it when they hear it. Smiling will enable you to defuse their hostility and allow them to hear you when you provide a reasonable solution to the problem at hand," Joe concluded.

"Yesterday we had a breakthrough," Joe continued. "The CEO's visit was more than just a recognition of the importance of our jobs here. It represents a sharing of responsibility. Our job is not just to sit here disconnected from the rest of the company and share sympathy but few real solutions with our customers. We are now connecting to the rest of the organization, and we will partner with them in solving customers' problems. We must adapt our roles to become agents of change. Our goal must be to change the way the company relates to customers and to change the way customers think about the company. It is only when those two things happen together that we will manage to facilitate a truly open relationship with customers. A relationship that will be based on mutual respect and true listening to each other. If we will adapt ourselves to become agents of change, then this company has a real chance of wining our customers' loyalty and business. I need your help to make this change happen. And all this started with a simple idea from Tim. I would like to thank Tim for his contribution."

They all cheered for Tim. He was one of them, and he had made an impact. They all were proud of him.

As Joe was walking the floor, he noticed that Dawn was engaged in a conversation with a customer. He waved to her and she motioned for him to come over. As he came closer, he was able to listen to the call. She was once again fully engaged, talking to the customer with complete willingness to resolve the problem as soon as possible. It was a pleasure listening to her. Joe was proud to have her as an employee.

After her call was finished, Dawn said, "Nice move bringing the executives to the call center. The visit was well received by the agents here."

"Well, thank you," Joe replied. "Let's have a coffee in the lunch room during your next break," Joe suggested.

"Sounds like a plan to me," Dawn replied with a smile.

Later, in the lunchroom, Joe asked Dawn about the progress in the call center and about how people were responding to the changes.

"Are you talking again about that software you installed?" she joked. "I think we established already that it was not such a great idea to impose it on people."

"No, not the software, though I still think it has great merit. It's a shame we're not using it to make our work easier and more fun. Other companies are taking full advantage of it and reaping great benefits from it," Joe responded.

"If that's the case, then how come none of us knows about it? It's just that . . ." she hesitated.

"It's just that what?" Joe prompted. "Really, it's okay to talk to me—I want to hear what you think."

"Well, installing the software may have been a really good idea. But I wonder if things would have gone better if we agents had been involved from the beginning. Since we really had never been told what the benefits were supposed to be, we simply disregarded the software."

"Do you expect me to go to every agent and ask their permission to put new software into the center? I'll never be able to make any decision if I try to reach that kind of consensus," Joe said.

"You seem to have reached consensus about the new principles. The working sessions we have with our supervisors generate a great deal of acceptance and move people toward adopting the principles. It really helps that you've started treating us as partners," Dawn said. "After all, can you imagine the company imposing a new product on the customers and forcing them to buy it? What do you think the customer reaction would be?" she asked.

"They would simply reject it and many might leave and go to the competitors," Joe replied, starting to get the point.

"Just as you want us to do with customers, it makes a huge difference for the better when companies treat their employees with respect, rather than imposing things on them against their will. When companies respect employees' opinions, employees are more likely to cooperate. Both customers and employees are human beings, and they share the same characteristics. If employees feel respected and included in the process, they'll be more likely to react favorably to almost any initiative," Dawn continued.

"I agree with what you're saying, but I still cannot wait for a consensus about every decision. Doing so would lead to chaos and anarchy," Joe resisted.

"I can see your point. Life is not always a democracy, and when it's not, your key to success, as a manager, is the trust you have built with your people and the way you communicate your decisions. Trust is something you build through your actions. Communication is critical. You're clearly building trust with your new approach, so you probably won't have a problem convincing them that your decisions are for their benefit, even if they were not involved in making those decisions," Dawn said.

"That's easier said than done," Joe replied, "because some decisions are not so easy to explain."

"I agree with you, but you're a great communicator. It all starts with your attitude: with understanding that you must communicate clearly and honestly with people. Honesty will take you very far in building trust," she concluded.

"Thank you for the advice. I sure have some food for thought," Joe said appreciatively.

"Anytime," Dawn replied, "but now I have to rush to pick up my child. Have a great day."

As he watched her walk away, another principle dawned on him, so he hurriedly wrote it on his napkin so he wouldn't forget it:

Earning customer loyalty starts with loyalty to employees.

All the way home, Joe reflected on this principle. It made a lot of sense. He must find ways to show employees that he cared about their work. After all, it isn't easy to spend all day talking to upset customers who demand things you can't give. He remembered when he started his career as an agent; there were days

Earning customer loyalty starts with loyalty to employees.

when he simply couldn't take the pressure anymore. Some customers were just too rude, and could be impossible to handle. He remembered how sometimes he dreaded the next call, not knowing who would "attack" him next. He must find a way to help the agents release their tension.

The next morning, the supervisors were all in the conference room waiting for the day's briefing. They had all been agents before becoming supervisors, so Joe figured they would be able to help him design the right program to further establish loyalty and trust.

"Today, I want to discuss the worst experience you had with customers when you worked here as an agent," Joe said.

They all looked at each other in amazement. Joe's morning meetings were becoming more interesting and surprising in the last two weeks. From yelling at them for poor results and threatening them, he had moved to asking them questions and listening to their answers. This transformation had an impact on them. They were starting to like the new Joe.

"This time I want to start with my own example. As you know, I started my career at this company as an agent; to tell you the truth, I remember those days with mixed feelings. Some days I could have just screamed at the customers. I often thought that they were rude and obnoxious. On days like that I thought the customers were there to cause trouble; it seemed as if being annoying was their sole purpose in life. It seemed as though they could always find a way to blame our company, regardless of the real cause of the problem, and they would always look for a way to squeeze something out of us.

But my worst feelings didn't come from handling these upset customers. It came from not being able to help those who were right and needed quick resolution. I often felt helpless, powerless to do anything but provide sympathy and read pre-prepared text. One day I got a phone call from a customer in New York. She was mad as hell and was ready to kill the next person who answered the phone. I guess the first agent who picked up her call sensed the anger, and immediately placed her on hold. The next agent claimed it wasn't his department and told her she would be transferred to an expert. While listening to our elevator music on the phone, she waited more than ten minutes before she actually reached the expert agent—me.

Of course, I was no expert, just an excuse for the other agents not to do their jobs. She started yelling and swearing at me, blaming the company for all of her problems. I had had a particularly bad day and did not have much patience for her. Still, I had no choice but to try to help her. Apparently she bought a discontinued product from one of the outlet stores. The product was faulty and she wanted her money back. The store's manager told her that no refunds were available for discontinued products, and that got her more upset. She argued that the policy should only apply to working products, not to faulty ones.

She had a point. If I had been her, I would probably have been upset too. I explained to her that, following procedure, I would send her a special form with three copies to fill out. Then she could officially submit her request to a committee. That answer got her even more upset. She wanted her money back, not a bunch of forms, and she refused to accept anything less. I tried to calm her down, but she refused to be satisfied with less than she felt she deserved, and she demanded that I not put her on hold or transfer her call elsewhere," Joe said.

"So what did you end up doing?" Kevin asked.

"I did the only thing I could to get out of the mess," Joe said. "I promised her a full refund and asked her to send the product to my attention. I didn't know where I would get the money, but I knew I had to get it."

"Wow! What a story!" Mark said. "I think you should share it with all the agents."

"Well can you imagine how I felt after this call?" Joe asked. They all nodded in sympathy. "I was wasted. I needed to vent; I needed to let off steam; I needed to hit something."

"So what did you do?" Mark asked.

"I took the next call, because I had nowhere to go. I don't think my story is unique. All of our agents are sometimes the focus of extraordinary anger and frustration. What may feel at the time like an abusive call really isn't. Customers are not bad people. They are people like you and me. They're just human and occasionally get emotional. We all do from time to time."

They all shared some of their horror stories, then Joe concluded, "I want ideas for ways to relieve the stress around here. We need to provide a place or an activity that allows our people to reenergize themselves and relieve their tension."

"How are you going to pull it off, with our budget freeze?" Ann asked.

"I'll justify it by claiming that the agents' productivity and quality of work will improve. You'll have to help me in that. Go out there and talk to your people. Have them share their horror stories, then ask them what would help them relieve their stress. Let's meet here tomorrow to discuss the ideas you've generated and see what we want to do about it," Joe added.

The thought that they might identify solutions and then be able to implement them was an idea they liked. They all left the room thinking about what might ease the pressure their agents felt, and eager to talk to them about it. They were going to love this assignment.

Joan was excited about her new idea and couldn't wait for the meeting to share it with her staff. Mark was skeptical about the possibility of getting the funds for the project, but he figured that if the CEO had already visited their call center, then anything was possible.

"Please close your eyes and try to remember the worst customer interaction you had in the last two weeks," Joan started. Her team was already getting used to these sessions, in which management wasn't telling them what to do, but was actually asking them what they thought. It was a major change, and they liked it. It was clear this was not temporary, but was an evolving process. They all complied with her request and spent the next few minutes thinking.

"I want you to focus on the bad interaction you had and on what you felt about it," Joan continued. They opened their eyes. Joan asked them who would like to go first.

"I'll go first." It was Mike, a quiet, reserved agent—and the last person Joan thought would be suitable to discuss frustration or anger. No one ever saw him get upset or lose his temper.

"Go ahead, Mike. Tell us about your bad experience."

"I actually had several of them, so it was difficult to choose one. The one I'll talk about happened last Tuesday. I received a call from a customer in Texas, who complained about not being able to find our product locally. I explained to him that we would be happy to send him our catalog, but he didn't like that option, saying he didn't want to pay the shipping and handling fees. I suggested he use our Web site, since we don't charge shipping and handling fees there, but he said he had no computer at home. I asked him why it was important to him to shop at the store. He just replied that he was the customer, therefore he was right and did not have to answer my question. I located the store nearest to his house, but he claimed that it was too far and that our company should pay for his gas and mileage to the store. I told him that was not an option.

After ten minutes of going back and forth, he started to raise his voice and yell at me. He threatened to go and buy a competing product. I asked him if our competitor had a store in his town, but he refused to answer and just continued to scream and swear. At that point I just told him that I couldn't help him and asked to finish the conversation. He refused to end the conversation and demanded to know my name. It was just a horrible experience, with no point to it."

"How did you feel?" Joan probed.

"I felt angry and upset. It put me in a bad mood. He had no right to speak to me like that. I'm not sure what his problem was, but he shouldn't have taken it out on me."

"I agree," Joan said. "Then what did you do?"

"Well, nothing, I just took the next call," Mike replied.

"Did it affect your performance for the rest of the day, or were you able to forget it and move on?" Joan asked.

"It pretty much ruined the rest of my day. I really didn't want to talk to anyone after that call. I know I was not at my best for the remainder of the day. Other customers suffered because of this guy."

"And if you had it your way, what would have released your stress after that horrible call?" Joan asked.

"I guess if I could've just gone and done something

else to take my mind off that call, it would have helped," Mike said.

"Give me an example," Joan requested.

"I guess a couple of jokes would have helped, or maybe a punching bag. Yes, actually, that would do; I really wanted to punch something or someone after that call." Everyone laughed but no one dismissed the idea.

The rest of the group all got a chance to tell their horrible interaction stories. They shared ideas for relieving stress. Joan listed them and promised to share them with Joe and the rest of the supervisors:

- A fun place full of games to play
- Joke Centers spread throughout the workplace
- A fitness room
- A Web site with jokes and inspiring thoughts
- A place where agents could share their frustrations with colleagues
- Cookie trays and candy jars

At the evening session, the supervisors shared the ideas they had gathered. They were all great ideas and Joe decided to implement several of them right away. Everyone agreed that no one was more suitable to design the Fun Center than Mark. Joan was assigned to work with IT to build a Web site that would enable the agents to share thoughts with each other. Kevin took responsibility for developing a "Best Joke of the Week" program, with joke boards spread throughout the workplace.

"Time is of the essence here," Joe said. "We must show our people that we take their ideas seriously and act on them quickly. They have entrusted their ideas to us, so we should show them the respect they deserve. If we want them to give their best and provide the best experience consistently, we must deliver similar experiences to them. We ought to show them that we are there for them and that their work is important. Let's make this happen. I'm sure that any means we can provide for reducing stress will increase our overall performance and reduce the turnover rate dramatically."

Joe saw the excitement in the supervisors' eyes. They were finally part of making something happen, not just sitting there to stop the masses from attacking the company. They had become true believers. He was confident they were sharing that excitement with their people.

That's the beauty of excitement, thought Joe. When it spreads, it's like wildfire—nothing can stop it. Joe had to admit that he was excited as well.

"I want our company to become *the* choice for anyone looking for the best, most challenging place to work," Joe concluded. He knew he wanted it to be his place of choice as well.

Chapter 8

★

The next day, Joe noticed some excitement at the entrance. He was not expecting anything special today, but to his surprise, it was the CEO with a larger team of executives.

"We're here to staff the phones," the CEO declared. "As I promised, this is going to become a regular practice."

"Welcome back. You're always welcome here. Let's get going," Joe responded, pleasantly surprised. "And just to make this new change truly permanent, I'm going to change our recorded message today. Instead of saying, 'This call may be monitored for quality purposes,' from now on it will say, 'This call may be answered by the CEO.' I think that will send a clear message to our customers about the fundamental change we're spearheading here."

That's a wonderful idea," the CEO said. "Let's get started, we have customers waiting." Joe and his supervisors were excited. There could not be a stronger message to their agents about the importance of their jobs than having the CEO perform that job. He could see the amazement in the agents' eyes.

During lunch, the executives ate with the agents in the cafeteria, exchanging thoughts and stories about the morning calls.

"I got a call today from an upset customer who damaged his clothing using our faulty product. He demanded that the company pay for his torn shirt," the CEO shared with the group eating around the table.

"Well, what did you do about it?" asked one of the agents.

"I told him to send the bill for a new shirt to my office, so I could reimburse him for it," said the CEO triumphantly. "He loved it. He told me he would continue to be our customer for many years. I feel so great," he concluded. To his surprise, only the executive team appreciated his story. The agents seemed a bit gloomy and unimpressed.

"Isn't that an example of great service and excellent customer experience?" he asked, surprised.

"Sure it is," the agent replied, clearly unimpressed. Something in his response bothered the CEO. The agent sounded disappointed rather than pleased.

"Then what's wrong?" the CEO asked.

"It's the kind of service only you are authorized to provide," the agent responded. "None of us can deliver similar service. We're not allowed to suggest that the company reimburse customers for their damage. That's beyond our authority."

"Then how would you respond to such a case?" the CEO asked.

"The official procedure is to tell them that the company is not responsible for misuse of the product. We often say this to people we know deserve compensation, but for whom we have no authority to provide it. Then they just get more frustrated," the agent replied.

"Then what do they do?" the CEO asked.

"They either threaten us, humiliate us, or simply declare that they will move their business to the competition. It's the most awful feeling to know that even if you do your best, they'll probably end up disappointed and leave the company. It feels as if our best is not good enough for them and we simply feel helpless," the agent concluded.

"Well, this has to change immediately. I'll take it up with Joe," the CEO said.

The CEO approached Joe's office. Joe looked at him, sensing that something was wrong. He knew that having the CEO back in the center so soon seemed too good to be true.

"Joe, we have a problem here," the CEO declared.

I knew it, Joe thought to himself, this whole thing was too good to be true.

"What can I do to help?" Joe asked.

"Actually, you can't do anything about it, it's my problem," the CEO said. Now Joe was confused.

"What is the problem?" he asked.

"The problem is that our people here are not empowered to solve customers' problems. They're here to talk to customers, but they have no power to resolve problems. How can we ask them to own the problem if we provide them no tools to solve the problem. This must be a very frustrating situation to be in," the CEO said. Joe could not believe what he was hearing. The CEO approached the white board and wrote on it:

Empower people
to solve problems.

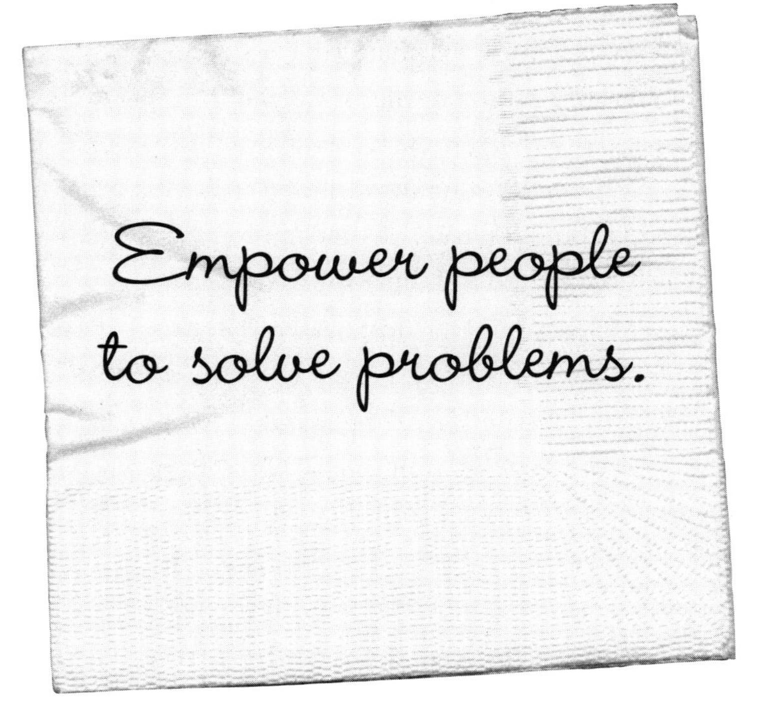
Empower people
to solve problems.

"This is a missing principle in your program," the CEO said. "Starting today, everyone will have authority—at his or her discretion—to spend up to $500 per call, without getting preapproval, to resolve customer problems, or to deliver an "extra mile" gesture. Anything over $500 and up to $1000 can be approved by their supervisors, and an expenditure of over $1000 can be approved by you. I want you to launch a training program to explain when it's appropriate to use this discretionary fund. I want to make sure people are not just here to chat with customers, but to solve their problems right away."

Joe was shocked. This was too good to be true. The CEO was truly getting it, and this wasn't just a temporary fix for him.

"I suggest you share the new plan with all the agents today. It's important that they hear it from the CEO—but I have another suggestion for you." Joe felt that since he was on a roll, he might as well push his new idea.

"I'll announce the program. What's your suggestion?" the CEO asked.

"It seems that you fully realize the power behind the call center. You see that nothing can replace the first-hand experience of talking to customers," Joe said.

"I agree with you completely," the CEO replied.

"Then let's try something new. All new recruits in the company, no matter what department they're hired in, should spend their first month here in the center talking to customers. I truly believe that this will be the best training they can get to work in this company. They will get firsthand insight of customers' thoughts and wishes. Because of their experience here, they will be better able later to apply and adapt their work to customer needs. By implementing this suggestion, you'll ensure that all employees are on the same page and are fully customer focused." Joe was afraid that he was pushing his luck. He already had a great deal of support from the CEO, and this suggestion could be one request too many. Still, the more he thought about it, the more it made sense. It had the potential of making the whole organization customer focused. He decided it was worth the risk of asking. He watched the CEO's face for his response.

The CEO's face didn't move for what seemed like an eternity. Joe was afraid he had really gone too far this time.

"I like it," the CEO declared, "and just to keep them all tuned up, each employee will staff the center one day every quarter," he concluded with a big smile.

"This is one of the best and simplest ideas I've heard for a long time. When someone comes to me with an idea, it usually carries a minimal impact and a large price tag. There's always a catch. This idea I can implement with great confidence, regardless of the additional costs, because it will have real impact on the organization. It really is a brilliant idea." Joe smiled and allowed himself to breathe a bit easier. His idea was worth the risk not only because it had been accepted, but also because he believed in it.

Later that day, all the agents were gathered in the meeting room when the CEO announced the new program. It was the first time that Joe had heard his people cheer, and the first time he had seen them acting upbeat. They were sharing high-fives and seemed quite excited. Looking at them, he knew they were all stepping into a new era.

"You should be proud of yourself," said Dawn, smiling. "Even in my wildest dreams, I did not think the center would be transformed so deeply. You have done an amazing job here. Congratulations."

"Thank you, but we all deserve the kudos. We all did it together. There's no way I could have done it alone. Everyone shared suggestions and we all changed the atmosphere around here to encourage positive attitudes; we all took charge of the situation. We all did it together. We all made our own experience better," Joe replied.

Chapter 9

On Monday, Joe decided to call Amy, the salesperson at CustomerUno. He invited her to conduct a sales pitch at his center.

"But you guys already bought our system!" Amy exclaimed, puzzled.

"Well, I did buy it," Joe replied, "but I was the *only* one here who bought it. I've got many people who need to be sold in order for us to use it and get the most out of it."

The following day, Joe split the agents into two groups. Each group was invited at different times to the meeting room for a demonstration and sales pitch of CustomerUno.

"You're here because I made a mistake and I want to fix it," Joe began. They listened to Joe very carefully. They had learned to appreciate his leadership, noticing the change for better that he was spearheading in their center. He was the reason their lives at the center had gotten better. They wanted to support more of his initiatives.

"A few weeks ago I decided, on my own, to purchase a new technology for the center. It was a mistake. Since you are the users, I should have consulted you first. That's what I want to do today. Please take the time to listen and ask any questions you have. At the end of the session we will conduct a vote. If the majority of you vote to use the system, then we will all do so."

Amy moved on to explain the benefits that the system could provide them in their daily tasks. She described how the software could save them time and effort, and allow them to do their jobs better. They all asked questions about different features of the program. Better still, they all seemed to like what they were seeing and seemed to approve of the system.

"Let's move to the vote," Joe proposed.

"Well, before we do so, I have one question," Dawn asked. "Can we customize the system so that it will include our guiding principles on the screen? It would be a great reminder of the principles we all believe in and act upon as a team." The others seem to like the idea.

Amy, who was already uncomfortable selling to the agents, looked quizzically at Joe. Joe nodded to her as if to approve the request.

"I guess we can do that, though this is the first time I've ever heard such a request," Amy replied. "But then, this is the first time I've sold to the actual users of our system, so I guess nothing should surprise me."

"Then, if the majority vote for CustomerUno, we will add the principles to the system," Joe concluded.

The vote passed unanimously. Joe felt as if it was more than just a vote about the system. It was a vote of confidence. He had won his employees' trust, and after all, that was more important to him than the system. They all seemed excited about CustomerUno, and ready to take full advantage of it. They got excited at some of the features that Amy presented, and asked questions that revealed the same excitement Joe had felt when he first heard about CustomerUno.

"It seems as if you've made some major changes around here," Amy said. "I remember the last time I was here, the place looked like a graveyard. Now people are excited and interested in what they're doing."

"When I bought CustomerUno, the organization was not ready for it. I couldn't force them to use it, and they rejected it out of hand. I needed to transform the people in order to make them ready to deliver the best possible customer experience," Joe said. "One thing I did learn is that people care for people, and technology is just a tool. I wanted to believe that there was a quick fix, but what seemed to be a shortcut proved to be the longest path to where I needed to go. It was also your biggest mistake. You should never portray CustomerUno as a product that delivers excellent service. No technology does. It's a great tool, but without the people behind it ready and willing to use it, it's useless. Yet *with* the cooperation of people, your software is priceless."

Chapter 10

Recently, Joe had started to staff the phones himself. He decided that this was a much better way of gathering statistics than reading all the reports that were piling up on his desk. Talking to customers and mingling with the agents provided him with more valuable information than any report could. He also felt that he was really contributing to the company's success and not stuck in a merely administrative role. Mark told him that he was now a leading role model, and that the agents had more respect for him and for what he said. He definitely noticed that change. Last weekend, he saw some of the agents in the park. They approached him and chatted with him, then introduced him to their families. It was gratifying. He felt a connection. It was a new experience for all of them.

The Joke Center was becoming a reality. Many jokes had been posted there. The agents now had a chat room in which they could share ideas and frustrations and provide support for each other. Joe joined the chat room often to see what everyone was talking about. He also provided some suggestions from his own experience. He was amazed at how quickly the agents adapted to the newly installed chat technology, as if they had been doing it for years.

"I just had a very frustrating call with a customer who would simply not listen to any reasonable explanation. This time, the customer *wasn't* right," wrote one of the agents.

"I know how it feels, but don't let it bother you too much. He probably had a bad day or was already upset before he called, and needed to let off steam," replied another agent.

"You're so good at reminding me when I forget that it isn't personal. It sure helps to talk to each other, though, when the customers are unreasonable. It helps us put things in perspective," replied yet another.

"Focus on those customers you actually did make happy today. It will lift your spirits," came another reply.

"Call the customer back in couple of hours and show him that you care. I'm sure he'll be more reasonable," was another response.

"Thanks, guys. You're all right. That's all I needed right now. Let's get back to work. We have customers who need us," wrote the original agent.

Joe was very proud of them. They truly cared for each other and for the work. He knew that morale was picking up. He had recently gotten a call from the recruiter the center used. The recruiter was concerned about a drop in business; the center was not hiring as much as they had in months past. He wanted to know what was wrong. Joe was smiling. Nothing was wrong. Actually, everything was right. Fewer people were quitting, so they didn't need to replace so many people.

The light was blinking on the phone set, signaling a waiting call.

"This is Joe, how can I help you?"

"I have an idea to help you improve your product. I recently used it in a new way and found it very effective. I think you guys should test it for this type of usage, and if I'm right, you should market it for that usage. It could open up a brand new market for you." The customer sounded excited about her discovery. Joe was caught up in her enthusiasm.

"I truly thank you for the input. Thank you for taking the time to call us," Joe replied courteously.

"What do you mean, 'thank you'?" the customer started to sound a bit annoyed. "What are you going to do with my idea? Is it going anywhere in your company?"

"Well, I'm not sure what to do with your idea. In our new customer relationship management system, we have no room for customers' ideas." Joe realized how stupid he sounded, even while he was still speaking.

"Well, in my vendor relationship management system, there is plenty of room for firing vendors who don't listen. You were just courteous with me, but you really didn't mean to do anything with my idea. What kind of a relationship is this?" the customer asked, clearly more annoyed. "If you truly mean to have a relationship with your customers, it must be on an equal basis. If you want my loyalty, you need to grant me a level of respect that shows me you deserve my loyalty. It's time to change the old format of relationship—you sell, we buy. If you want a relationship with me, you had better not just listen in order to placate me. You'd better listen in order to act upon what I'm telling you. I want a part in shaping the products, not just in buying them and increasing your revenues." The customer had obviously thought about these ideas earlier—they were not just a burst of anger. She made sense. She was right, but he did not know what to do with the suggestion.

"You're right. I agree with you. I wouldn't want to be 'managed' or 'handled' if I were a customer. I promise you I will personally take your idea to the CEO. You'll hear from me soon about what we did with your idea. In the meantime, I'd like to offer you a $100 gift certificate for taking the time to share your idea with me."

"That's more like what I expected from you," the customer responded with an audible smile. "Thank you for taking the time to truly listen."

Reflecting on what had just happened, Joe realized the next important principle:

Listen to learn.

He realized that he was programmed to listen in order to respond, solve the problem, and make it go away. He was not listening simply to hear the customer. He never had any intention of doing something with what he was hearing. Even if he heard a new idea, he wouldn't know where to send it. His agents were probably facing similar calls every day and had no idea what to do with their customers' suggestions. The cen-

Listen to learn.

ter and his agents were an isolated island, disconnected from the rest of the business.

Come to think of it, mused Joe, the whole system of measuring and compensating the agents was based on who answered the most calls, in the least amount of time. In such an environment, agents would not want to listen, because it would make their calls last longer. So, in fact, the more his agents actually listened to their customers, the less they earned. Joe realized that the transformation he had started was far from complete. He could only imagine the frustration it caused agents who heard good ideas but were powerless to do anything with them. It must be equally frustrating to choose between listening to customers or earning their bonuses. In the current system, the two were mutually exclusive. Joe knew it was time for a more fundamental change.

At the next morning meeting he shared his new discovery with his supervisors. "It dawned on me that our system is geared toward efficiency; we actually pay our agents based on their efficiency. The problem is that, at the same time, we claim to care about building and sustaining relationships with our customers. Last I checked, those two goals—efficiency and relationship building—cannot coexist. You cannot have an "efficient relationship." You either invest time in building a relationship or you let go of the relationship and become efficient. Can you imagine being "efficient" with your loved ones? I can't imagine being efficient with my wife. Our relationship grows through shared experiences, and time is key to this growth. The longer we're together, the better our relationship becomes as we invest time in learning to understand each other. We're sending mixed messages to our agents and to our customers. Customers are told through our brochures and advertising that we seek a sincere and committed relationship, which takes time to develop, while our agents are being paid *not* to spend time with customers. This simply does not make sense!" Joe declared.

They all nodded in agreement. They looked relieved.

"As far as I'm concerned, if a customer wants to talk to us, we should listen, regardless of the length of the call. We can always learn something new," Joe said. "Even upset customers can teach us about what we can improve on. Upset customers have options. If they want to stop doing business with our company, they can send us a letter to announce their decision. They can simply take their business elsewhere without saying a word. If they've bothered to call, they're still trying to salvage the relationship—otherwise they wouldn't have spent the time calling us. So if they call us, we ought to listen. Their call means we still have a chance to make it right for them. They sure are giving us another chance by calling. Why shouldn't we give them the same chance?" Joe concluded.

"Starting today," declared Joe, "every agent who comes up with a new idea, after talking to a customer, will receive a special bonus. Additionally, I will work with human resources to adjust the compensation plan so that agents' performance will be measured—and they will be compensated—according to our true goal: to deliver the best experience to our customers. We want existing customers to have an experience that will make them return, and we want new customers to leave the competition and stay with us," Joe finished with pride.

Everyone around him smiled in agreement. It was a new era in the center. It was a major change, and they recognized it as such, and were happy to be part of it.

"But then what happens when an agent comes up with a new idea? Where does it go?" Mark asked.

"Excellent point, Mark," Joe responded. "We'll create a list of the different product managers, support managers, manufacturing managers, accounting managers, and any other manager to whom customers' ideas or suggestions may be relevant. This list will be given to the agents. I want them to send the ideas directly to the right manager. Five days later, they will follow up with the manager to check the status of the idea, and then call the customer back to let him or her know what progress has been made. That way, we'll be able to demonstrate to our customers that we truly listen and intend to act on their suggestions. Our customers will realize that we're here to treat them as equal partners in this business."

"Isn't it too dangerous for the agents to go directly to our top managers?" Joan asked.

"If we trust our agents with our most important assets in this company—our customers and their loyalty—then why can't we trust them to talk to our executives?" Joe replied. "It's important that they feel their efforts are being directly recognized and that no supervisor is stealing the show from them. I believe that it will be consistent with our principle of being loyal to our people and treating them with respect. I have no doubt that they will act responsibly and not abuse the privilege," Joe concluded.

They all seemed to agree. They left the room to inform their agents of the latest change.

By mid-afternoon, Joe had twenty new ideas in his mailbox. It seemed as if the agents had taken the challenge seriously and had started to talk to customers with the intention to learn. He was amazed at how quickly they adapted. In the past, any new rule took weeks, if not months, to gain full adherence, but this one was gaining compliance much faster.

Joe looked at his computer screen and saw the principles all flashing in front of him, in the reminder function of the CustomerUno software they all had decided to buy.

Technology does not care for customers, people do.

Treat the customer the way you want to be treated.

Own the problem.

The customer is everyone's job!

Ensuring customer loyalty starts with loyalty to employees.

Empower people to solve problems.

Listen to learn.

It had been quite a journey to discover the principles, but they were all right on target. This is a people business, first and foremost, Joe realized.

As he stepped out of his office to look at the agents in action, he noticed the electronic boards on the wall. The old measurement of "Number of Calls Waiting" had been replaced. The lights on the board were now indicating "Number of Ideas Generated," "Number of Ideas Implemented," across the company and the "Savings and Revenue" that those ideas generated for the company. By now, many departments of the company regularly visited the center. Employees all came to listen and talk to customers. New recruits spent their first weeks at the center getting to know the company and its products from the customers' point of view. Product managers came regularly to speak to customers about new ideas for products. The center became truly central to the company. All these great results had been achieved by his people. He was proud of them.

Yet he did not need the electronic boards to tell him what was obvious. Customer experience improved dramatically because his people cared. And they cared because he cared about them. It was one of those things that you don't need scientific proof to accept. You know it when you see it. The improvement, however, didn't have to be taken on faith alone—the results were concrete. As customer satisfaction increased dramatically, the revenues from repeat sales to those customers increased as well. It was a true measure of the customers' appreciation of the service level his people delivered. His people's morale was up and that meant a dramatic reduction in turnover, which resulted in a precipitous drop in hiring and training costs. These savings he invested in stress reduction programs, such as the Joke Center. What he also noticed, to his pleasure and surprise, was that his own work experience improved. He no longer dreaded going to work, but rather came to work with the energy and the willingness to make a difference. It seemed as if the old ideals had come back to life. He was living the life he wanted, and he was making an impact. The employees' experiences were creating a customer experience that affected his personal experience—in one beautiful circle.

Epilogue

★

"Joe, it's Joan; I have an escalated call. This customer demanded to speak to you and no one else." It had been a while since he had needed to handle an upset customer. The number of escalated calls had reduced so dramatically that the supervisors were able to resolve the problems and appease the customers. He had almost forgotten how to handle these calls.

"This is Joe, how can I help you?" he asked, dreading what would come next.

"Am I speaking to the call center manager?" asked the voice on the other end of the line.

"Yes, this is Joe Jacobs and I'm the manager of this center." Joe noticed that he was saying it with a touch of pride. It was a new thing for him.

"Well, I wanted to tell you that I've been a customer of this company for more than ten years, and for most of those years, you delivered the same consistently lousy service. It was embarrassing. When I called your center today, I said to myself, 'This is it. This is the last time I deal with this company.' I was ready to quit and switch to the competition—but I was amazed by the change. I was answered by a knowledgeable person willing to help, who after hearing the history of my problem, went out of her way to offer compensation. And on top of that, she asked me if I had any suggestions to improve your products or services. No one ever asked me for my opinion before." The customer spoke with excitement. "I don't know what you did to your people, but whatever you're doing, you're on the right track. I'm staying with your company. This is the kind of company that I want to give my business to. Thank you for an excellent experience. You guys made my day. I just have one last question, though."

"What is it?" Joe replied, smiling.

"What will you surprise me with tomorrow?"

"Just watch," Joe answered, though not sure what he really meant.

"I will," the customer said. "Again, thank you for making my day."

"What will you surprise me with tomorrow?" The question echoed in Joe's mind. What did the customer mean by that? He did like our service, so what was wrong with it? Why did he want another surprise? Isn't it better if we keep things the way they are? Reflecting throughout the day on what the customer had said, he decided to share it with Dawn during their coffee break.

"It sounds simple to me," Dawn said. "Let me give you an example. If you're in a relationship with someone and every day you go to the same restaurant and see the same TV shows and read the same books, what would you feel about the relationship after a while?"

"It would get boring," Joe answered.

"You got it. Customers are human and they get bored."

"Don't you think they'd prefer consistency?" Joe asked.

"They'll expect us to consistently deliver superior service, but they also expect excitement. They also look for ways to rejuvenate the relationship, so they'll remember why they're in it. Remember how you told us to relate to our personal relationships so we can understand the customers better?" Dawn asked.

"Of course," Joe replied. "Doing so allows us to relate to the relationships with customers as relationships with human beings."

"That's right. And just as we must rejuvenate our personal relationships and take them to new heights, keeping customers excited—not bored—is a good way to keep them away from the competition. If customers are excited, they'll have no reason to look around for alternatives."

"So I guess this customer was not really demanding anything from us, but rather giving us the best secret of keeping him with us. He was helping us figure out how to retain his loyalty," Joe said.

"It sure sounds that way," Dawn replied, "but remember, the excitement should never come at the expense of the quality service that customers already expect. It has to be something extra."

Just as Joe thought he'd figured out all the principles of *The Experience*, he realized that the most important one of all was missing. Experience cannot stay constant. It requires rejuvenation and renewal, just as every human relationship does. After all, thought Joe, that's what a customer relationship is—a human relationship. It seemed as if all the other principles had led to this one. The others formed the foundation to start a relationship with the customer but didn't provide the momentum to keep it going. The experience must constantly improve. The company must deliver a new, exciting experience every day, so that customers never get a chance to think about the competition. Come to think of it, mused Joe, the same rule applied to his own employees. They, too, need *The Experience* of daily rejuvenation and reinvention to stay motivated and excited.

Keep them excited.

Reflecting on this newly discovered principle, Joe was happy that he had not taken the promotion the CEO had offered him. He'd been given the opportunity to manage a large products division, but he liked it at the center. He joined the center in order to make people happy, and he was living his own dream. He liked to keep close to the customers. In his job at the call center, he had the chance every day to make an impact and to create a great experience for other people and for himself. And he continually learned new ideas from customers and from his own agents. They gave him his best experiences.

Joe realized that keeping customers excited would keep him challenged for quite some time. His job was just beginning. He needed to find ways for his team to reinvent the relationship and the experience they delivered every day. And he had to do that without

Keep them excited.

compromising the level of service his team had already achieved. As we reinvent our customer relationships, things can only get better. It could be fun, he thought to himself. Joe knew that he was facing the beginning of a journey full of challenges, a journey he would need to pursue every day; but he also knew that this journey would be an exciting personal experience.

The Experience!

How to Wow Your Customers and Create a Passionate Workplace

Technology does not care for customers, people do.

People come first. Technology can help. It is a tool; what people do with it, is up to you.

Treat the customer the way you want to be treated.

Being able to relate to our personal experiences as customers is key to our ability to deliver the best experiences to our customers.

Own the problem.

If you will not own it, no one will. Personal responsibility is key to delivering customer experiences.

The customer is everyone's job!

Customer service is not a department, it is the job of the whole organization. Everyone who works at the company is responsible for delivering the best customer service through their respective roles.

Earning customer loyalty starts with loyalty to employees.

Customer experience is about human interaction. If you want your people to deliver the best experience, give them a good reason to do it. The best reason is delivering to your people the best experiences.

Empower people to solve problems.

Do not put people in a position of having responsibility without authority. Give them the tools to solve the problems they encounter. If you trust them to take the call and handle your customers, trust them to resolve the problems.

Listen to learn.

Recognize that your customers know your product as well as you do. Stop listening in order to answer and start listening like a student who is ready and willing to receive valuable information. Act on customer suggestions to show your customers that you care enough to build a relationship based on mutuality.

Keep them excited.

As in personal relationships, consistency can only go so far. Customers are people and they take for granted what you delivered yesterday. Reinvent the relationship and rejuvenate the experience. Make them excited and keep them coming back for more.

With Gratitude

This book is the culmination of many experiences of many people, including my own. It is through those experiences that the principles of customer experience management were discovered. I would like to thank all the people who delivered bad as well as good experiences, and through those experiences, led me to the writing of this book.

Special thanks to Bill Wear for his editing insight, to Matt Kelsey and Frank Brogan for believing in this book and making it a reality. To Jacqueline Dever for editing and making the final touches that made the difference. My deepest gratitude belongs to my Drora, my wife and my true friend, for all the support and encouragement throughout the years. It is your love and support that keeps me going and gives me an experience of a lifetime. To my family for enduring my passion and actually finding a place in you to love it. Thank God.